The Wit and Wisdom of Polar Animals

This is a STAR FIRE book

STAR FIRE BOOKS
Crabtree Hall, Crabtree Lane
Fulham, London SW6 6TY
United Kingdom

www.star-fire.co.uk

First published 2007

07 09 11 10 08

1 3 5 7 9 10 8 6 4 2

Star Fire is part of The Foundry Creative Media Company Limited

The CIP record for this book is available from the British Library.

ISBN: 978 1 84451 807 4

Printed in China

Thanks to: Cat Emslie, Andy Frostick, Sara Robson,
Gemma Walters and Nick Wells

The Wit and Wisdom of Polar Animals

Ulysses Brave

enlightenment

Foreword

For years I studied Zen and the Art of
Animal Self-consciousness. Subsequently I
have written a large number of management,
self-help and philosophical texts over the
years, which have provided helpful advice
to those less fortunate than myself.
Here then, is my latest offering.

Ulysses Brave

At times of great stress
it is often best to remain still
for some considerable time.

The path to enlightenment can be accompanied by significant risk.

If you study your own reflection too intensely, you will fall into its seductive arms.

*If you fall, consider
remaining in this new position.
It will provide a fresh
perspective to your
familiar environment.*

Try to avoid the rush hour if you feel depressed. Loneliness is most acute when in the company of others.

*Try to avoid the obvious
photo opportunities, they
rarely capture the inner self.*

It is easy to be misunderstood while you focus on your Chi.

When dealing with
depression it is important
to have a series of strategies
which cope with even
the most mundane of
everyday situations.

Relatives can help you put your life into perspective.

Before combat, focus your energies. Your solar plexus will vibrate with Chi and you will be unstoppable.

Try not to feel inadequate if your chosen method of disguise lets you down.

Backwards leaping is like one-handed clapping. For the Zen master, it is second nature. For the novice, much practice is required.

*Do not retreat into denial —
dreams and fantasies have
their place, but not in
everyday living.*

A cool assessment of your emotional relationships can result in deep and sudden loneliness.

Leadership requires massive self belief. Sometimes those in the lead are, in fact, at the back end of a giant circle.

When attempting to climb the career ladder, try plunging into new challenges without preparing for the consequences.

*Grace and power are rare
bedfellows so do not expect
miracles at first.*

Don't shy away from affection, but be wary of strangers with Big Love.

*If you find yourself in an
uncomfortable social situation,
always call a style consutant.*

Beware of your past. It can break through your conscious defences at any time and blow feelings of guilt and shame across your calm, happy exterior.

*If you look in the mirror
and hate what you see, try
to remove all mirrors
from your home.*

Negotiation is important in love-making. Try to get as close as possible to achieve the maximum result.

Try not to resort to artificial stimulants. The real world has plenty of life-amplifying plants.

Sometimes, it is best not to look around or behind because the presence of others can be disconcerting.

*Try to set yourself
realistic goals.*

Ancient signs and
symbols remain important.
Always maintain a high
level of discipline when
giving directions.

Group consciousness can be a powerful tool in a community which knows how to wield its power.

Try to find a safe place to consider the solutions to your personal problems.

Even the tallest or largest amongst us have the basic needs of emotional comfort and recognition.

The inner self is often more attractive than the outer appearance.

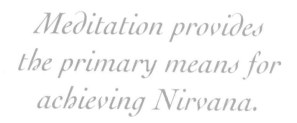

*Meditation provides
the primary means for
achieving Nirvana.*

A few moments of preparation before a major event is essential.

Young love can be explosive if allowed to get out of hand.